NUREG-1837

Regulatory Effectiveness Assessment of Generic Issue 43 and Generic Letter 88-14

Manuscript Completed: September 2005
Date Published: October 2005

Prepared by:
J.V. Kauffman, NRC Project Manager

Division of Systems Analysis and Regulatory Effectiveness
Office of Nuclear Regulatory Research
U.S. Nuclear Regulatory Commission
Washington, DC 20555-0001

ABSTRACT

This report documents an assessment of Generic Issue 43, "Contamination of Instrument Air Lines," and Generic Letter 88-14, "Instrument Air Supply System Problems Affecting Safety-Related Equipment." This assessment is part of an ongoing initiative through which the U.S. Nuclear Regulatory Commission (NRC), Office of Nuclear Regulatory Research, is reviewing selected agency regulations and decisions to determine whether they are achieving the desired results. For this assessment, the staff compared expectations with outcomes. Whenever outcomes fell short of expectations, the staff attempted to identify ways to enhance the effectiveness, efficiency, and realism of the NRC's regulatory process.

On the basis of its assessment, the staff concluded that licensee and agency activities, such as the Maintenance Rule, Generic Letter 88-14, design-basis reconstitution, and others, have significantly improved air system and component performance and, thereby, resulted in improved reactor safety. Moreover, issuance of Generic Letter 88-14 and targeted NRC inspections led to the identification and resolution of air system design issues impacting safety-related systems and components, again resulting in improved reactor safety. As a result, based on data for pressurized-water reactors, major losses of instrument air are now infrequent, and prompt recovery from such losses is typical, which supports the staff's conclusion that reactor safety has improved. In addition, as evidenced by the ongoing discovery and correction of air system issues, licensee programs and NRC oversight activities provide assurance that the NRC and its licensees are effectively maintaining reactor safety in this area.

FOREWORD

The U.S. Nuclear Regulatory Commission (NRC), Office of Nuclear Regulatory Research (RES), is reviewing selected agency regulations and regulatory actions to determine whether they are achieving the desired results. This review is part of an evolving initiative to make NRC activities and decisions more effective, efficient, and realistic, in accordance with the NRC's Strategic Plan for Fiscal Years 2004–2009 (NUREG-1614, Volume 3, dated August 2004).

For the particular review discussed in this report, the staff's goal was to assess the effectiveness of Generic Issue 43, "Contamination of Instrument Air Lines," and Generic Letter 88-14, "Instrument Air Supply System Problems Affecting Safety-Related Equipment," by comparing their stated expectations with actual outcomes. In conducting this assessment, the staff's primary source of outcomes was data concerning actual experience in operating the Nation's nuclear power plants. To glean that operating experience data, the RES staff reviewed licensee event reports, inspection findings, and summary analyses of operating experience, such as initiating events studies and studies of the reliability of air systems and their components. Whenever outcomes fell short of expectations, the RES staff attempted to understand how and why this occurred, and to identify possible ways to enhance the regulatory process. In addition, in 2005, the NRC staff distributed a draft of this report for internal peer review. The RES staff subsequently considered and addressed the resultant comments, as appropriate, in preparing this final report.

On the basis of its assessment, the staff concluded that licensee and agency activities, such as the Maintenance Rule, Generic Letter 88-14, design-basis reconstitution, and others, have significantly improved air system and component performance and, thereby, resulted in improved reactor safety. Moreover, issuance of Generic Letter 88-14 and targeted NRC inspections led to the identification and resolution of air system design issues impacting safety-related systems and components, again resulting in improved reactor safety. As a result, based on data for pressurized-water reactors, major losses of instrument air are now infrequent, and prompt recovery from such losses is typical, which supports the staff's conclusion that reactor safety has improved. In addition, as evidenced by the ongoing discovery and correction of air system issues, licensee programs and NRC oversight activities provide assurance that the NRC and its licensees are effectively maintaining reactor safety in this area.

The RES staff will continue to conduct long-term reviews of operating experience in order to identify safety issues and opportunities to improve regulatory effectiveness, efficiency, and realism.

Carl J. Paperiello, Director
Office of Nuclear Regulatory Research
U.S. Nuclear Regulatory Commission

CONTENTS

Figures

EXECUTIVE SUMMARY

The U.S. Nuclear Regulatory Commission (NRC), Office of Nuclear Regulatory Research (RES), is reviewing selected agency regulations and regulatory actions to determine whether they are achieving the desired results. This review is part of an evolving initiative to make NRC activities and decisions more effective, efficient, and realistic, in accordance with the NRC's Strategic Plan for Fiscal Years 2004–2009 (NUREG-1614, Volume 3, dated August 2004).

The goal of this review, as it relates to this report, is to assess the effectiveness of Generic Issue 43, "Contamination of Instrument Air Lines," and Generic Letter 88-14, "Instrument Air Supply System Problems Affecting Safety-Related Equipment," by comparing their stated expectations with actual outcomes. In conducting this assessment, the RES staff's primary source of outcomes was actual operating experience. To glean operating experience data, the staff reviewed licensee event reports, inspection findings, and summary analyses of operating experience, such as initiating events studies and studies of the reliability of air systems and their components. In particular, the staff sampled licensees' submittals in response to Generic Letter 88-14, and related agency actions, as well as inspection reports and safety evaluations.

The NRC staff initiated Generic Issue 43 in response to an event at Rancho Seco Nuclear Generating Station in July 1981. The staff subsequently reevaluated the issue as part of a priority evaluation in 1983, and recommended dropping it from further consideration. However, following the publication of the priority evaluation in November 1983, the staff received comments from the NRC's Advisory Committee on Reactor Safeguards (ACRS) and Office for Analysis and Evaluation of Operational Data (AEOD). Rather than agreeing with the staff's recommendation, the ACRS and AEOD comments recommended that the staff should broaden the issue to include all causes of air system unavailability, instead of confining it with the restrictive limits that were previously imposed in Generic Issue 43.

The NRC's Office of Nuclear Reactor Regulation (NRR) concurred with the ACRS and AEOD recommendations and agreed to reevaluate the issue following the completion of an extensive AEOD case study of air systems. The AEOD staff completed Case Study C/701 in March 1987 and subsequently published NUREG-1275, "Operating Experience Feedback Report — Air System Problems," Vol. 2, dated December 1987. As a result of that case study, the NRC staff reevaluated and broadened Generic Issue 43, and subsequently assigned the issue a high priority ranking based on its value/impact score. To alert licensees and industry stakeholders to the broader issue concerning instrument air supply system problems affecting safety-related equipment, the staff then issued Generic Letter 88-14 on August 8, 1988; the staff considered Generic Issue 43 resolved with the issuance of Generic Letter 88-14.

The staff issued Information Notice 2002-29, "Recent Design Problems in Safety Functions of Pneumatic Systems," October 15, 2002 (Ref. 7), to inform addressees of recent occasions where the controls or design of safety-related systems incorporating non-safety-related air-operated controls was less than adequate. NRC also expressed a long-standing concern for such, often subtle problems; and discussed prior missed opportunities to identify them.

Now, more than 2 decades after Generic Issue 43 first arose, the staff has evaluated its effectiveness, including the related Generic Letter 88-14, by comparing stated expectations with actual outcomes.

On the basis of its assessment, the staff reached the following conclusions:

- Licensee and agency activities, such as the Maintenance Rule, Generic Letter 88-14, design-basis reconstitution, and others, have significantly improved air system and component performance and, thereby, resulted in improved reactor safety.

- Based on data for pressurized-water reactors, major losses of instrument air are now infrequent, and prompt recovery from such losses is typical, indicating that the actions requested by Generic Letter 88-14 have contributed to improved reactor safety.

- Issuance of Generic Letter 88-14 and targeted NRC inspections led to the identification and resolution of air system design issues impacting safety-related systems and components, and resulted in improved reactor safety.

- As evidenced by the ongoing discovery and correction of air system issues, licensee programs and NRC oversight activities provide assurance that the NRC and its licensees are effectively maintaining reactor safety in this area.

ABBREVIATIONS

AC   alternating current
ACRS   Advisory Committee on Reactor Safeguards (NRC)
ADAMS   Agencywide Documents Access and Management System
AEOD   Analysis and Evaluation of Operational Data, Office of (NRC)
AFW   auxiliary feedwater
AOV   air-operated valve
ASP   accident sequence precursor

BWR   boiling-water reactor

CA   control air
CCW   component cooling water
CFR   *Code of Federal Regulations*
CS   core spray
CSB   containment spray building
CVCS   chemical and volume control system

DC   direct current
DG   diesel generator

EA   Enforcement Action
ECCS   emergency core cooling system
EDG   emergency diesel generator
EOP   emergency operating procedure
EPIX   Equipment Performance and Information Exchange
EPRI   Electric Power Research Institute
ESW   emergency service water

FCV   flow control valve

HPCI   high-pressure coolant injection
HPSW   high-pressure service water

IA   instrument air
IAS   instrument air system
INEEL   Idaho National Engineering and Environmental Laboratory
ISI   inservice inspection
IST   inservice testing

LER   licensee event report
LOCA   loss-of-coolant accident
LPSI   low-pressure safety injection
LWR   light-water reactor

MDAFWP   motor-driven auxiliary feedwater pump

NRC   U.S. Nuclear Regulatory Commission
NRR   Nuclear Reactor Regulation, Office of (NRC)
NUDOCS Nuclear Documents System

PA   plant air
PORV   power-operated relief valve
PRA   probabilistic risk assessment
PWR   pressurized-water reactor

RBBCW   reactor building closed cooling water
RCIC   reactor core isolation cooling
RES   Nuclear Regulatory Research, Office of (NRC)
RHR   residual heat removal

SCSS   Sequence Coding and Search System
SOV   solenoid-operated valve

TDAFWP turbine-driven auxiliary feedwater pump

U.S.   United States (of America)
UFSAR   updated final safety analysis report

1. INTRODUCTION

The U.S. Nuclear Regulatory Commission (NRC), Office of Nuclear Regulatory Research (RES), is reviewing selected agency regulations and regulatory actions to determine whether they are achieving the desired results. This review is part of an evolving initiative to make NRC activities and decisions more effective, efficient, and realistic, in accordance with the NRC's Strategic Plan for Fiscal Years 2004–2009 (NUREG-1614, Volume 3, dated August 2004).

The goal of this review, as it relates to this report, is to assess the effectiveness of Generic Issue 43, "Contamination of Instrument Air Lines," and Generic Letter 88-14, "Instrument Air Supply System Problems Affecting Safety-Related Equipment," by comparing their stated expectations with actual outcomes. In conducting this assessment, the RES staff's primary source of outcomes was actual operating experience. To glean operating experience data, the staff reviewed licensee event reports, inspection findings, and summary analyses of operating experience, such as initiating events studies and studies of the reliability of air systems and their components. In particular, the staff sampled licensees' submittals in response to Generic Letter 88-14, and related agency actions, as well as inspection reports and safety evaluations.

The NRC staff initiated Generic Issue 43 in response to an immediate action memorandum issued by the NRC's Office for Analysis and Evaluation of Operational Data (AEOD) in September 1981. Specifically, that immediate action memorandum was prompted by an incident at Rancho Seco Nuclear Generating Station in July 1981, in which the presence of desiccant particles in a valve operator resulted in slow closure of a containment isolation valve. Consequently, the AEOD memorandum concerned the common cause failure potential associated with desiccant contamination of instrument air lines (Ref. 1).

The NRC's Office of Nuclear Reactor Regulation (NRR) responded to the AEOD memorandum by establishing a working group to determine the generic implications of air system contamination and to develop appropriate recommendations (Ref. 2). Desiccant contamination of the plant instrument air system (IAS) was also one of the contributing causes of the loss of the salt water cooling system at San Onofre Nuclear Generating Station in March 1980, which caused the staff to issue Generic Issue 44, "Failure of Saltwater Cooling System." However, since the only new generic concern identified in the evaluation of the San Onofre event was the common cause failure of safety-related components as a result of IAS contamination, the staff subsequently combined Generic Issue 44 with Generic Issue 43 (Ref. 3).

The staff subsequently evaluated Generic Issue 43, and recommended dropping it from further consideration (Ref. 3). However, following the publication of the priority evaluation in November 1983, the staff received comments from the NRC's Advisory Committee on Reactor Safeguards (ACRS) and AEOD. Rather than agreeing with the staff's recommendation, the ACRS and AEOD comments recommended that the staff should broaden the issue to include all causes of air system unavailability, instead of confining it with the restrictive limits that were previously imposed in Generic Issue 43.

The NRR staff concurred with the ACRS and AEOD recommendations and agreed to reevaluate the issue following the completion of an extensive AEOD case study of air systems at light-water reactors (LWRs) in the United States (Ref. 3). The AEOD staff completed Case Study C/701 in March 1987 and subsequently published NUREG-1275, "Operating Experience Feedback Report — Air System Problems," Vol. 2, dated December 1987 (Ref. 4). As a result of that case study, the NRC staff reevaluated, broadened, and retitled Generic Issue 43. The retitled Generic Issue 43, "Reliability of Air Systems," was subsequently assigned a high priority ranking based on its value/impact score.

To alert licensees and industry stakeholders to the publication of the AEOD case study in NUREG-1275, Volume 2, the staff issued Information Notice 87-28, "Air Systems Problems at U.S. Light Water Reactors," Supplement 1, December 28, 1987 (Ref. 5). The staff then issued Generic Letter 88-14 on August 8, 1988 (Ref. 6), to alert licensees and industry stakeholders to the broader issue concerning instrument air supply system problems affecting safety-related equipment. The staff considered Generic Issue 43 resolved with the issuance of Generic Letter 88-14.

The staff issued Information Notice 2002-29, "Recent Design Problems in Safety Functions of Pneumatic Systems," October 15, 2002 (Ref. 7), to inform addressees of recent occasions where the controls or design of safety-related systems incorporating non-safety-related air-operated controls was less than adequate. NRC also expressed a long-standing concern for such, often subtle problems; and discussed prior missed opportunities to identify them.

Now, more than 2 decades after Generic Issue 43 first arose, the staff has evaluated its effectiveness, including the related Generic Letter 88-14, by comparing stated expectations with actual outcomes.

2. BACKGROUND

2.1 General Background of Air Systems

In publishing NUREG-1275, Volume 2 (Ref. 4), AEOD provided detailed background information and a description of air systems. This section summarizes material from Sections 1, 2, and 3 of that report. Note that this is an abbreviated, generalized description of air systems, which vary significantly in design from plant to plant. Although a more complete list appears in NUREG-1275, Volume 2, the following important equipment and systems typically use instrument air:

- scram system
- reactor coolant system (pump seals and relief valves)
- safety injection system
- auxiliary feedwater system
- primary containment isolation system
- chemical and volume control system, charging and letdown system, and boration system
- high-pressure injection system/makeup system
- automatic depressurization system
- low-temperature overpressurization protection system
- component cooling water system
- decay heat removal system
- service water system
- emergency diesel generators
- reactor cavity, spent fuel, and fuel handling system
- torus and drywell/vent and vacuum system
- station batteries
- main steam system, main steam isolation valves, and auxiliary boiler
- reactor building/auxiliary building ventilation and isolation system
- main feedwater system and main feedwater isolation valves
- standby gas treatment system

Many LWRs in the United States rely upon air systems to actuate or control safety-related equipment during normal operation. However, at most LWRs, the air systems themselves are not classified as safety systems. Plant safety analyses typically assume that nonsafety-related air systems become inoperable during transients and accidents, and the air-operated equipment fails in known, predictable modes (e.g., fails open, fails closed, or fails as is). In addition, air-operated equipment and systems that must function during transients or accidents are provided with a backup air or nitrogen supply in the form of safety-grade accumulators.

Most LWRs have several air systems. In general, the highest purity air system, typically referred to as instrument air (IA) or control air (CA) is used for vital instrumentation and controls. Most LWRs also have lower quality air systems, which are typically referred to as plant air (PA), service air, or station air. These lower quality air systems are commonly used for nonsafety-related equipment, routine maintenance, pneumatic tools, and breathing air. As such, they are usually allowed to operate with larger particulates and higher moisture and oil content than IA systems.

Typical air systems are made up of two or more 100-percent capacity compressors that deliver air pressure at about 100 psig. When the IA system pressure decreases below a predetermined setpoint (typically about 75 psig), the redundant air compressor automatically starts and the main air header sheds the PA system. Many plants have other backup air sources, such as portable skid-mounted diesel- or gas-driven compressors. At some plants, the backup air supply is of relatively low quality and may feed directly into the IA system downstream of the air dryers and filters. When such backup sources are operating, the potential for contaminating the IA system can be significantly increased.

An air line that penetrates containment is usually equipped with an automatic isolation valve, which closes on a containment isolation signal. Some plants have a separate air system to supply air-operated equipment inside containment. These separate air systems have some advantages. For example, air supply inside containment is not necessarily lost upon containment isolation, and a malfunctioning or leaking IA system inside containment does not result in a containment pressure increase because the system draws upon the containment atmosphere for its supply.

2.2 Background and History of Generic Issue 43

Reference 3 contains a detailed history of Generic Issue 43. This section summarizes material from Reference 3, with additions from referenced documents as noted.

As previously stated, the NRC staff initiated Generic Issue 43 in response to an immediate action memorandum issued by AEOD in September 1981. Specifically, that memorandum was prompted by an incident at Rancho Seco Nuclear Generating Station on July 7, 1981, in which the presence of desiccant particles in a valve operator resulted in slow closure of a containment isolation valve. AEOD characterized the desiccant contamination as a potential common cause failure of pneumatically operated equipment, which threatened about 130 safety-related items. Consequently, AEOD recommended taking immediate actions at Rancho Seco and issuing an appropriate bulletin. In addition, AEOD wanted licensees to furnish a listing of their experience with air system contamination, provide an assessment of the safety implications of those events, and evaluate their plants' susceptibility to contamination-induced common cause failure of the air system (Ref. 1).

The NRR staff responded to the AEOD memorandum by establishing a working group to determine the generic implications of air system contamination and to develop appropriate recommendations (Ref. 2). The NRR staff also stated that the NRC was evaluating maintenance at Rancho Seco, and any identified problems would be corrected before restarting the plant.

Staff review revealed that desiccant contamination of the plant's IAS was also one of the contributing causes of the loss of the salt water cooling system at San Onofre Nuclear Generating Station in March 1980, which caused the staff to issue Generic Issue 44, "Failure of Saltwater Cooling System." However, since the only new generic concern identified in the evaluation of the San Onofre event was the common cause failure of safety-related components as a result of IAS contamination, the staff subsequently combined Generic Issue 44 with Generic Issue 43 (Ref. 3).

The staff subsequently prioritized Generic Issue 43 and recommended dropping it from further consideration (Ref. 3). However, following the publication of the priority evaluation in November 1983, the staff received comments from ACRS and AEOD. Rather than agreeing with the staff's recommendation, the ACRS and AEOD comments recommended that the staff should broaden the issue to include all causes of air system unavailability, instead of confining it with the restrictive limits that were previously imposed in Generic Issue 43. The NRR staff concurred with the ACRS and AEOD recommendations and agreed to reevaluate the issue following the completion of an extensive AEOD case study of air systems at LWRs in the United States (Ref. 3).

The AEOD staff completed Case Study C/701 in March 1987 and subsequently published NUREG-1275, "Operating Experience Feedback Report — Air System Problems," Vol. 2, dated December 1987 (Ref. 4). In so doing, AEOD considered the multitude of events in which degraded or malfunctioning air systems had adversely affected safety systems, and viewed them as important precursor events. AEOD's primary concern was the potential for common mode failures that could result in the simultaneous loss of required safety systems. AEOD concluded that some plants with significant IAS degradation might be operating (or might have operated) with much higher risk than previously estimated. AEOD did not have high confidence that licensees would voluntarily take corrective action to avoid plant operation with degraded air systems in the absence of a serious event, because many plants do not have specific license requirements prohibiting operation with degraded IA systems. Consequently, AEOD recommended initiating the following actions either by the industry or through the regulatory process (Ref. 4):

- Ensure that air system quality is consistent with equipment specifications and is periodically monitored and tested by the licensee.

- Review the adequacy of anticipated transient and system recovery procedures and related training for loss of air systems, and revise as necessary.

- Train plant staff regarding the importance of air systems.

- Verify the adequacy of safety-grade backup air accumulators for safety-related equipment.

- Require all operating plants to perform tests involving a gradual loss of IAS pressure.

As a result of the AEOD case study, the NRC staff reevaluated, broadened, and retitled Generic Issue 43, and subsequently assigned the issue a high priority ranking based on its value/impact score. In so doing, the staff recognized that this high priority was driven by the analysis of risk attributable to IA failure at a single plant (Oconee Nuclear Station, Unit 3). That analysis revealed a high degree of sensitivity to IA failures, which was primarily attributable to poor selection of a "fail safe" position as a result of a loss of operating air for one particular valve (which was later changed). Accordingly, the analysis used as a surrogate for all plants was very plant-specific in nature. However, past licensee event reports (LERs) related to air systems revealed numerous additional instances in which a high degree of risk sensitivity was apparent. Therefore, the staff used the probabilistic risk assessment (PRA) for Oconee 3, as modified by Brookhaven National Laboratory, in order to ascertain an industry-wide risk estimate, recognizing that it would not be appropriate for all plants and was no longer appropriate for Oconee 3.

To alert licensees and industry stakeholders to the publication of the AEOD case study in NUREG-1275, Volume 2, the staff issued Information Notice 87-28, "Air Systems Problems at U.S. Light Water Reactors," Supplement 1, December 28, 1987 (Ref. 5). The staff then issued Generic Letter 88-14 on August 8, 1988 (Ref. 6), to alert licensees and industry stakeholders to the broader issue concerning instrument air supply system problems affecting safety-related equipment. The staff considered Generic Issue 43 resolved with the issuance of Generic Letter 88-14.

2.3 Generic Letter 88-14

Generic Letter 88-14 asked each licensee and applicant to review NUREG-1275, Vol. 2, and perform a design and operations verification of the IAS, which was to include the following considerations:

(1) verification by test that actual IA quality is consistent with manufacturer's recommendations for individual components served.

(2) verification that maintenance practices, emergency procedures, and training are adequate to ensure that safety-related equipment will function as intended on loss of IA.

(3) verification that the design of the entire IAS, including air or other pneumatic accumulators, is in accordance with its intended function, including verification by test that air-operated safety-related components will perform as expected in accordance with all design-basis events, including a loss of the normal IAS. This design verification should have included an analysis of current air-operated component failure positions to verify that they are correct to ensure required safety functions.

In addition, Generic Letter 88-14 asked each licensee and applicant to provide a discussion of its program for maintaining proper IA quality. Licensees and applicants were to provide their submittals within 180 days, signed under oath or affirmation indicating that the licensee or applicant had completed the requested actions or provided its plan and schedule to complete the requested actions. In addition, Generic Letter 88-14 also stated that each submittal should identify any components that cannot accomplish their intended safety function, and should state the corrective action taken or to be taken. In addition, Generic Letter 88-14 asked each licensee and applicant to provide written notification after completing all actions, and to retain documentation from the verification for 2 years for future audit by the staff.

2.4 Expected Results from the Resolution of Generic Issue 43 and Issuance of Generic Letter 88-14

The actions requested in Generic Letter 88-14 express the NRC's expectations regarding this issue. Very simply, the NRC expected licensees and applicants to verify and ensure that their safety-related air-operated equipment and components were operable and could perform their intended design functions. This effort was to ensure adequate air quality and also included design verification and verification by test that equipment would perform as expected for all design events, including a loss of normal IA. A key aspect of the verification was to ensure that component failure positions were correct to ensure required safety functions. The NRC also expected licensees and applicants to have a program to maintain proper

6

IA quality. Part of the verification was to ensure that maintenance, procedures, and training were adequate to ensure that safety-related equipment would function as intended on a loss of IA. Thus, the regulatory expectations were that licensees and applicants would identify and correct IA problems, and put programs in place to maintain the IA systems and air quality. As such, licensees and applicants would identify and correct design vulnerabilities, especially those that could result in common mode failures, and no (or only very few) failures of safety-related equipment attributable to IA problems would be expected after licensees and applicants completed the actions requested in the generic letter.

8

3. REGULATORY ASSESSMENT

For this regulatory effectiveness assessment, the RES staff compared the actual outcomes of Generic Issue 43 and Generic Letter 88-14 with their stated expectations. Whenever outcomes fell short of expectations, the staff attempted to understand how and why this occurred, and to identify possible ways to enhance the regulatory process. In conducting this assessment, the RES staff's primary source of outcomes was actual operating experience. To glean operating experience data, the staff reviewed LERs, inspection findings, and summary analyses of operating experience, such as initiating events studies and studies of the reliability of air systems and their components. In particular, the staff sampled licensees' submittals in response to Generic Letter 88-14, and related agency actions, as well as inspection reports and safety evaluations. The staff then compared the final outcomes of Generic Letter 88-14 to the issues and recommendations identified in NUREG-1275, Volume 2, to evaluate the extent to which those issues and recommendations were addressed in the final resolution of this topic.

3.1 Operating Experience — Recent Summary Studies

Recent summary studies of operating experience data show improved performance of air systems and their components since 1988.

3.1.1 Rates of Initiating Events at U.S. Nuclear Power Plants (1988–2003)

In 2004, Idaho National Engineering and Environment Laboratory (INEEL) analyzed the initiating event frequencies at U.S. nuclear power plants (Ref. 8). As show in Figures 1 and 2, the historical frequency of initiating events with loss of instrument air has decreased for both boiling-water reactors (BWRs) and pressurized-water reactors (PWRs). INEEL concluded that the long-term trend is likely to represent more than mere random variation. Both BWRs and PWRs showed low baseline initiating event frequencies for loss of instrument air, with mean frequencies of 0.0083 and 0.0115 per reactor critical year for BWRs and PWRs, respectively. These baseline initiating event frequencies are significantly lower than the "pre-baseline" frequencies (before 1994 for BWRs and before 1990 for PWRs); however, the INEEL analysis did not provide reasons for the observed frequency changes.

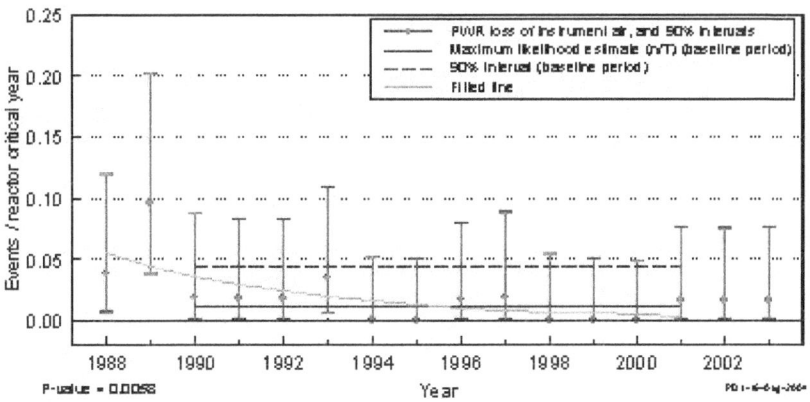

Figure 1. Frequency of PWR initiating events with loss of instrument air

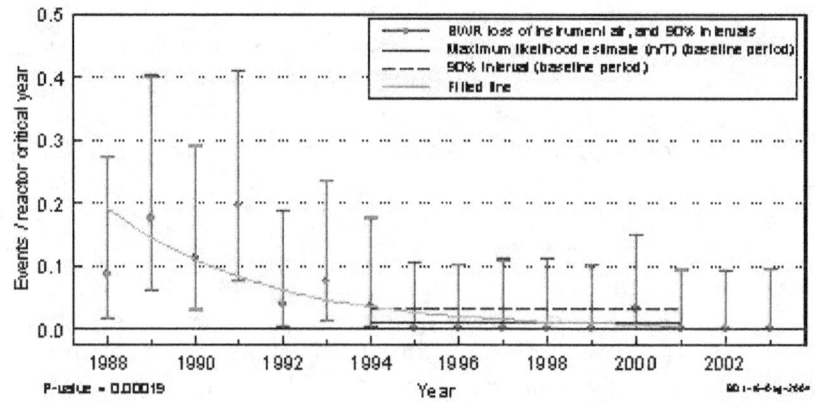

Figure 2. Frequency of BWR initiating events with loss of instrument air

3.1.2 Component Performance Study of Air-Operated Valves (1987–2003)

In 2004, INEEL also analyzed the component performance of air-operated valves (AOVs) at U.S. commercial reactors (Ref. 9). That analysis showed statistically significant improving trends for AOV performance in many areas, including overall failure frequency (Fig. 3), frequency of unplanned demands (Fig. 4), and probability of failure on demand (Fig. 5), each of which reduced by a factor of 3 or greater. The INEEL analysis did not provide reasons for the observed performance improvement.

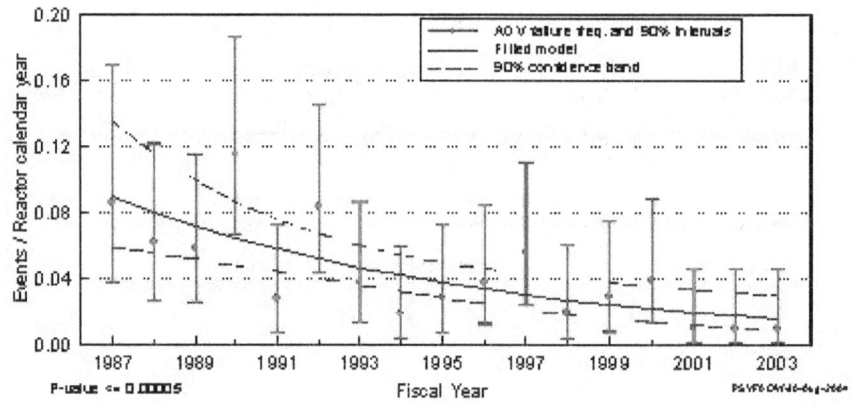

Figure 3. Frequency of failures (events per operating year), as a function of fiscal year

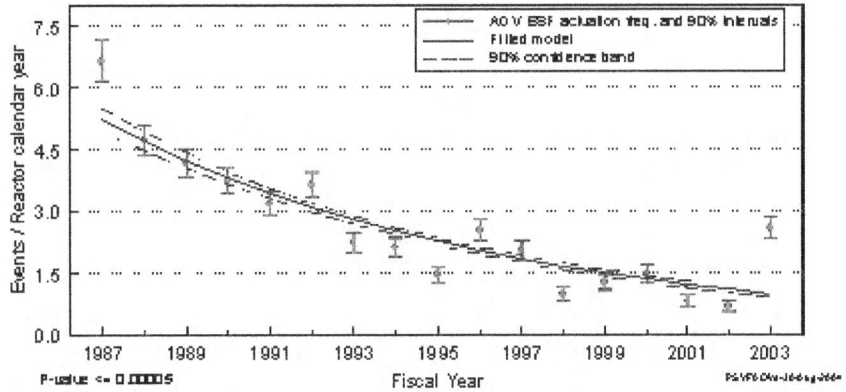

Figure 4. Frequency of unplanned demands (events per operating year), as a function of fiscal year

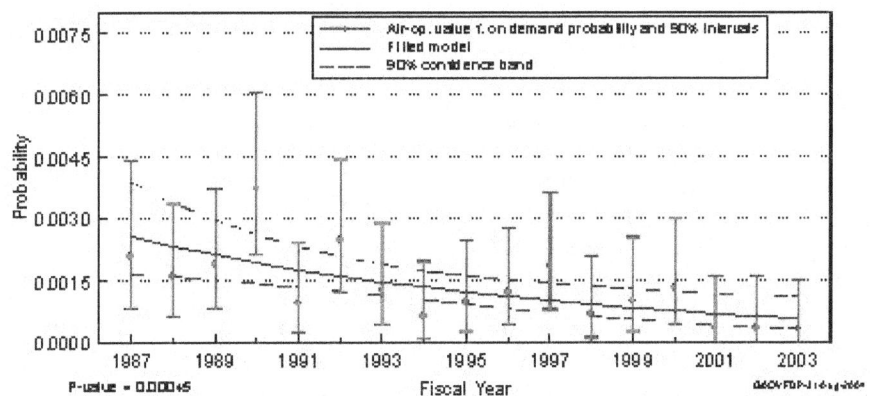

Figure 5. Probability of air-operated valve failures on demand

11

3.1.3 Recovery from Loss of Instrument Air Events

In the final accident sequence precursor (ASP) analysis for LER 266/01-005 and Enforcement Action (EA) 02-031 (see Section 3.3), the staff evaluated 20 industry events involving a total loss of instrument air at PWRs for the period from 1987 through 2001. In so doing, the staff's objective was to calculate the probability of failure to recover instrument air. Toward that end, the staff rejected relatively minor events for a variety of reasons (e.g., the event only involved a local loss of air pressure to selected components, instantaneous air recovery was attributable to automatic fault isolation, or loss of instrument air did not cause a reactor trip). In addition, other calculations accounted for the loss of electric power to the air compressors following a loss of offsite power. The ASP analysis revealed the following observations:

- Six events contributed to the "failure to recover" calculation for a total loss of instrument air.

- Four of the six events were recovered within 4 minutes of the reactor trip (LERs 400/87-041, 280/90-006, 285/90-026, and 306/96-002).

- The remaining two events were recovered within 30 minutes of a reactor trip (LERs 317/87-003 and 456/88-025).

In summary, major losses of IA are infrequent at PWRs. Prompt recovery from a major loss of IA is typical at PWRs.

3.1.4 Findings from Recent Summary Studies of Operating Experience

The staff's analysis revealed the following findings concerning recent operating experience:

- Instrument air system and component performance has significantly improved since 1987.

- Prompt recovery from a major loss of IA is typical at PWRs. All of the events that occurred after the issuance of Generic Letter 88-14 were recovered within 4 minutes, with the exception of the event reported in LER 456/88-025. Note that the event associated with LER 456/88-025 occurred only about 3 months after issuance of Generic Letter 88-14.

3.2 Operating Experience — Licensee Event Reports

The staff queried the Sequence Coding and Search System (SCSS) to identify reportable events or conditions related to air systems from 1988 through February 2004. (The staff did not query the SCSS for events after February 2004 because SCSS was discontinued in February 2004.) The following subsections describe some of the more important events or conditions and provide observations based on the binning of LERs.

3.2.1 Licensee Event Reports with Air Systems and Generic Letter 88-14

A search linking "air systems" and "Generic Letter 88-14" returned 26 LERs, including 20689003, 20689025, 20690006, 21989001, 21989008, 24796016, 25989003, 26990010, 27093002, 27189001, 27198021, 27790025, 29388021, 29389002, 31390010, 31789005, 31789018, 32490004, 32588034, 32792018, 33889002, 36989006, 36989007, 38909007, 42396040, and 45889024. The following paragraphs briefly describe the most significant LERs (i.e., those that impacted multiple trains or systems).

LER 20689003 reported that the licensee identified that temperature control valves that regulate component cooling water (CCW) flow to the residual heat removal system could fail open as a result of an assumed loss of instrument air. This would cause the CCW flow to the recirculation heat exchanger to decrease below that assumed in the safety analysis. In addition, if this scenario occurred coincident with a single failure that rendered only one CCW pump operable, the resultant configuration would create a potential CCW pump runout condition that could result in a total loss of CCW. The licensee concluded that the single failure analysis performed in 1976 did not include the effects of the failure of these temperature control valves and, consequently, the licensee did not recognize the susceptibility of the CCW and emergency core cooling systems (ECCSs) to failure of these valves.

LER 27790025 reported that the licensee identified a condition that could prevent the high-pressure service water (HPSW) and emergency service water (ESW) ventilation systems at both units from automatically operating during design-basis events involving a loss of IA. Specifically, the licensee discovered that the temperature control system could fail to operate as a result of unavailability of the IA supply to the control system during such design-basis events. The failure of the ventilation system to automatically start could potentially cause HPSW and ESW pump motors to overheat and fail. HPSW pumps provide heat removal for the containment cooling mode of the residual heat removal system. ESW pumps provide cooling to the emergency diesel generators (EDGs), core standby cooling system and reactor core isolation cooling (RCIC) system area compartments, core spray (CS) motor coolers, and residual heat removal (RHR) pump seal coolers. The licensee attributed the cause to a deficient design of the HPSW and ESW ventilation control system. The licensee attributed a contributing cause to a design review oversight associated with Generic Letter 88-14.

LER 31789018 reported that the licensee's testing identified a condition that could prevent the fulfillment of certain systems to remove residual heat and control the release of radioactive material after a loss-of-coolant accident (LOCA). Specifically, the licensee discovered that many air-operated valves and piston-operated ventilation dampers that use safety-related air accumulators would not perform as expected after a loss of normal nonsafety-related IA. Affected systems included ECCS pump room ventilation, spent fuel pool ventilation, EDG service water, auxiliary feedwater (AFW), and saltwater. The licensee attributed the root cause to the lack of an adequate documented design basis, combined with inadequacies in the testing and preventive maintenance programs for instrument air systems. The licensee also identified several contributing causes. One contributing cause was that the function of many IA components as post-accident mitigators was unclear and undocumented. This led to design errors during system modifications and, ultimately, to a difference between design and actual system configurations. Another contributing cause was that the original accumulator design did not account for leakage through the control valve actuators and regulators. Another contributing cause was that there was no periodic testing and maintenance program in place to ensure that the critical IAS design parameters were preserved. In addition to the poor maintenance practices, poor component selection led to some identified leakage. Finally, the licensee stated that maintenance on the nonsafety-related IAS was generally given a low priority based upon its designation.

Findings from LERs with Air Systems and Generic Letter 88-14

The staff's analysis revealed the following findings concerning the 26 LERs that involved air systems and referenced Generic Letter 88-14 since 1988:

- Issuance of Generic Letter 88-14 and targeted NRC inspections resulted in the identification of problems in safety-related systems and components by both the NRC and licensees.

- The LERs reported a wide variety of problems and programmatic issues, and often contained several related issues in a single LER.

- About 80 percent (21 of 26) of these LERs were reported before 1991 (i.e., within about 2 years after issuance of the generic letter). Half (13 of 26) of the LERs described previous missed opportunities to identify the reportable condition. Three of these LERs described the missed opportunities as including licensee activities associated with Generic Letter 88-14.

- About 65 percent of the conditions reported in the LERs were identified by the licensee, about 23 percent were self-revealing, and about 12 percent were identified by the NRC.

3.2.2 Licensee Event Reports with Air Systems and Loss of Safety Function

A search linking "air systems" and "loss of safety function" returned 128 LERs. A detailed reading of these 128 LERs identified 58 that involved degradation of at least one train of a safety system. These 58 LERs are 20689024, 21390016, 21393005, 24595010, 24788017, 24793010, 24797010, 25596001, 26697014, 26992012, 26994003, 27190013, 27598003, 27789006, 27790004, 28295009, 28588009, 28588010, 28590016, 28591019, 28592011, 28598008, 28696002, 29889010, 29893009, 29896015, 30297014, 30589005, 30996027, 31597026, 31598052, 32792015, 33197013, 33389004, 33491005, 33694039, 33696017, 33696020, 33698019, 33698025, 33892003, 34189005, 34199006, 34602004, 34603001, 34688007, 36988036, 37396019, 37493005, 41090021, 42396010, 42396028, 42396031, 42397013, 44094014, 45889022, 45896007, and 45898003. This section also includes a recent event notification. The following paragraphs briefly describe the most significant LERs (i.e., those that impacted multiple trains or systems).

LER 24788017 reported that as a result of questions raised during an NRC inspection, the licensee determined that a single failure could cause a loss of air supply to the pneumatic controls of the EDG building ventilation system. This loss of air supply, in turn, could disable the entire ventilation system and, under certain conditions, could render the EDGs inoperable. The EDG building ventilation system was originally designed in the 1960s.

LER 27790004 reported that after special testing, the licensee identified that the ESW system would not have performed its safety function under "worst case conditions" for design-basis events involving a loss of offsite power, which would cause the high-pressure coolant injection (HPCI) system, two CS pumps, one RHR pump, and RCIC to be inoperable. Specifically, for a design-basis event including a loss of offsite power, the service water system and IAS are assumed to be unavailable. Upon loss of air, the air-operated ECCS and RCIC pump room cooler isolation valves would fail open, and some heat removal would still occur for these rooms using ESW. The ESW system at this site is common to both units, and also provides cooling water to EDG heat exchangers. The licensee attributed the proximate cause to gradual buildup of corrosion products and silt on the interior wall of the ESW piping, which resulted in higher resistance to flow. The licensee's root cause analysis revealed several contributing causes, one of which was weakness in understanding ESW design bases (including the significance of a loss of IA on ESW). The duration of this condition for each unit was indeterminate.

LER 28588010 reported that testing revealed that check valves in IA lines for level instrumentation on the safety injection and refueling water tank failed to hold a back-pressure, as would be required after a loss of IA. If a loss-of-coolant accident (LOCA) occurred with a coincident loss of IA, it is possible that a recirculation actuation signal would have actuated earlier than designed, resulting in a loss of safety injection and containment spray flow as a result of a lack of water in the containment sump. The licensee replaced the check valves with a different type and incorporated them into the inservice inspection (ISI) program.

LER 29889010 reported that the NRC raised concerns regarding the failure position of certain diesel generator (DG) room ventilation system pneumatically controlled air dampers, cooling water valves, and heating steam valves during an inspection. Subsequently, the licensee determined that a failure of the non-essential air supply would cause the dampers and valves to fail, resulting in elevated temperatures in the DG rooms. These elevated temperatures could prevent the DGs from completing their safety function. The licensee attributed the root cause to an error in the original plant design, which used a non-essential, non-seismic air supply interface to safety-related ventilation components.

LER 30996027 reported that an NRC Independent Safety Assessment Team identified that the inlet vanes for ventilation fans that provide cooling for containment spray pumps and dual-purpose low-pressure safety injection (LPSI)/RHR pumps could fail shut on the loss of nonsafety-related IA (see EA-96-299). Loss of ventilation to the motors for these pumps could cause them to fail as a result of overheating. The licensee attributed the cause to a failure to consider the failure modes of the inlet vanes during original plant design.

LER 31597026 reported that the licensee recognized that there was the potential for common mode failure of both trains of safety-related equipment because of a lack of overpressure protection on the various control air headers, if an air regulator failed open and resulted in an overpressurization of a control air header. Overpressurization of the 20-psig header could result in degradation of the RHR system and partial opening of two steam generator power-operated relief valves (PORVs). The licensee attributed the lack of overpressure protection to the fact that a regulator failing open was not identified as a mechanism that could overpressurize the low-pressure air headers.

LER 31598052 reported that the licensee identified a failure of the nonsafety-related manual loader for the turbine-driven auxiliary feedwater pumps (TDAFWPs), which could result in the TDAFWPs operating at a minimum speed with only one motor-driven auxiliary feedwater pump (MDAFWP) in operation when AFW is required to mitigate the consequences of an accident. This could occur if the speed control manual loader fails such that the TDAFWP governor valve receives the compressed air header pressure of 20 psig. The licensee attributed this condition to a failure to consider all failure modes when the compressed air system was originally designed. The licensee also identified a similar failure mode for the RHR heat exchanger air-operated outlet valves and centrifugal charging pump discharge flow control valve.

LER 33389004 reported that the licensee identified a design deficiency that would result in a loss of area cooling for parts of both safety divisions of safety-related and nonsafety-related electrical distribution systems as a result of a loss of IA to the cooling system temperature control valves. Specifically, the valves would fail closed when they should fail open. This was an error in the original design. This design deficiency involved both safety divisions and could potentially affect more than one ECCS.

LER 33694039 reported that the licensee identified some components that might not be included in the inservice testing (IST) program. The component listing was quite cryptic (only valve designations were provided), but appeared to involve the charging, containment spray, feedwater, instrument air, station air, diesel generator, safety injection, service water, and other systems. The licensee attributed this condition to personnel error during IST program development.

LER 33696020 reported that the licensee identified that following certain design-basis events, both trains of the reactor building closed cooling water (RBCCW) system could become inoperable if inventory were lost through the common plant makeup water system fill line. The licensee concluded that the makeup supply line must be assumed to be depressurized during accident scenarios because the plant makeup pump is not powered by a vital bus. The makeup supply line isolation valve also fails open on loss of air, so isolation would not be maintained. This could divert RBCCW flow to the plant makeup water surge tank. Assuming this loss of inventory without makeup, it is postulated that both RBCCW trains would eventually be unable to provide heat removal during certain design-basis events. The licensee attributed the cause of this condition to an original design deficiency.

LER 34602004 reported that the licensee determined that the pressure-regulating valve setpoint for the reactor coolant pump seal injection valves was inadequate to ensure closure of the valves upon receipt of a containment isolation signal. This condition represented a potential common mode failure. The licensee stated that it was unable to identify a formal calculation supporting the design basis and appropriate actuator settings during original plant construction. The licensee also stated that the basis is unknown for the pressure setpoint of a modification in 1977 to add a pressure regulator. In addition, the licensee stated that the cause of the misorientation of the isolation valves appeared to stem from confusion or miscommunication during original installation of the valves.

LER 34603001 reported that the licensee determined that several AOV actuators had negative operating margins, and a total of eight valves were not capable of performing their safety functions for the most limiting conditions. The affected valves were in the component cooling and service water systems. The licensee stated several causes for this condition during original construction. Specifically, many AOV actuators were undersized for a variety of reasons, AOV actuators were sized with minimum built-in margin, and there were similar analytical deficiencies in the design of accumulators. The licensee stated that it was implementing an AOV reliability program.

LER 34688007 reported that the licensee discovered the potential for air-operated temperature control valves for the decay heat removal heat exchangers to move from their fail-safe position upon loss of IA following initiation of the safety features actuation system. The licensee made this discovery while evaluating a similar condition of the service water temperature control valves for the component cooling water system. The valve operators are equipped with two accumulators, but the design was such that air leakage from an accumulator could cause the valves to drift open.

LER 36988036 reported that a licensee audit identified two design deficiencies associated with the DG starting air system. The first problem involved the safety classification of the piping and equipment from the DG starting air compressors to the check valve at the starting air tank inlets, which indicated that makeup to the tanks was both nonsafety and non-seismic. The second problem involved improper isolation of the DG starting air system from the IAS blackout header, which was neither safety-related nor seismically qualified. The blackout header could have depressurized the DG starting air system during a seismic event and shut down the DGs. There were no calculations or tests to demonstrate that the DG starting air system could supply blackout air header components and DG control demands. Past operability of the DGs was not conclusively determined, although if adequate DG starting air pressure is not maintained, the DG fuel racks will position to the fuel-off position. The design deficiencies existed on both units since initial startup.

LER 42396031 reported that licensee review identified that 41 solenoid-operated valves (SOVs) that perform an active safety function were susceptible to excessive operating pressure differentials resulting from failures of non-qualified air regulators in the IA lines upstream of the SOVs. The affected systems included high- and low-pressure safety injection, chemical and volume control, AFW, charging, and main steam. The licensee attributed the cause of this condition to failure to consider the potential for pressure regulator failure in the original design and selection of SOVs.

LER 45898003 reported that licensee engineering personnel identified an error in the design of the DG trip system logic that would degrade the ability of the Division I and II DGs to perform their safety function (see EA-98-478). Specifically, the design error would cause an unintended DG trip if control air pressure is lost. The licensee attributed the root cause to a design error in the electro-pneumatic control logic of the DG trip control logic during original design. The effect of the error was that the non-essential trips that were intended to be bypassed during emergency operations would be re-activated as control air pressure decreased below 120 psig, and a DG trip would occur if control air pressure decreased to 40 psig. DGs would have been capable of performing their intended safety function with credit given for operator action.

EN 40724, dated May 4, 2004, reported that the licensee discovered an unanalyzed condition with respect to containment pressure following a main steam line break. In the scenario of interest, loss of offsite power results in the loss of power to the permanently installed plant instrument air compressors, so that IA pressure would begin to decay. A double-ended guillotine break of a main steam line normally assumed, but a smaller break slows the rate of pressure loss from the affected main steam line. Break sizes exist such that by the time the main steam system pressure decays to the AFW actuation setpoint, IA pressure might become inadequate to close the feedwater control valves, thereby allowing additional inventory to reach the steam generators and flash to steam. The licensee connected a diesel air compressor to the IA header. The diesel air compressor has been placed in operation to ensure a source of instrument air in the event of a loss of offsite power.

Findings from Licensee Event Reports with Air Systems and Loss of Safety Function

The staff's analysis revealed the following findings concerning the 58 LERs that involved air system design conditions and potential or actual degradation of a least one train of a safety system since 1988:

- Issuance of Generic Letter 88-14 resulted in the identification and correction of problems in many safety-related systems and components.

- The major safety systems involved (in order of decreasing frequency) were emergency power, injection and cooling systems, containment isolation, AFW, service water, containment spray, and component cooling water.

- About 28 percent of these LERs were reported before 1991 (i.e., within about 2 years after issuance of the generic letter), and some licensees reported LERs before the issuance of the generic letter. About 72 percent of these LERs were reported in 1991 or later, suggesting that initial licensee and NRC activities to identify and resolve reportable air system problems had only limited effectiveness.

- Most of these LERs involved long-standing conditions, with two-thirds (or more) existing since initial construction and licensing.

- About 80 percent of the conditions reported in the LERs were identified by the licensee, about 10 percent were identified by the NRC, and about 10 percent were self-revealing. About half of the NRC-identified conditions were in the 1988–1990 time frame, when the NRC was inspecting air systems and reviewing licensee actions in response to Generic Letter 88-14.

3.3 Operating Experience — Inspection Findings and Escalated Enforcement Actions

A search of the NRC's Agencywide Documents Access and Management System (ADAMS) linking titles containing "inspection finding" and document text containing "air" returned 91 matches. Of those 91 matches, only 4 actually involved inspection findings associated with air systems or components. In addition, a review of escalated enforcement actions since 1996 (available at http://www.nrc.gov/reading-rm/doc-collections/enforcement/actions/reactors/) identified 12 that involved air systems or components.

NRC Inspection Report No. 50-244/01-010: The NRC issued a non-cited violation for the licensee's failure to identify that the support for containment isolation valve AOV-966C did not meet the screening criteria for seismic qualification and, therefore, was not properly evaluated. The licensee declared the penetration inoperable and closed the redundant containment isolation valve pending resolution of the problem. This inspection finding was classified as green.

NRC Inspection Report No. 50-220/02-10; 50-410/02-10: The inspection team identified a lack of adequate corrective action to address long-standing problems with the Unit 2 IAS. Following an IAS modification in 1993, problems were identified with IA compressor cooling water pump trips and cycling, as well as the need for operator action to restart the IA compressors after a loss of offsite power, which could affect the reliability of the IAS. Although the problems were entered into the corrective action program, there was a history of canceled deviation event reports and long-standing operator "work-arounds" associated with the IAS. There was no violation of NRC requirements because the IAS was not safety-related. This inspection finding was classified as green.

NRC Inspection Report No.50-458/02-07: An issue that arose during this inspection became a white finding (EA-03-077). Specifically, in May 2002, a full-flow condensate filter bypass valve was improperly manipulated such that, if a large feedwater transient occurred, such as one that would occur following a reactor scram, the valve could fail closed and the feedwater and condensate systems would be lost as sources of makeup water to the reactor vessel. Such an event occurred on September 18, 2002, following a reactor scram. In that instance, the valve failed closed, resulting in a loss of condensate and feedwater systems.

The valve condition arose when the bypass valve was installed in a section of condensate pipe that handled full system flow without redundancy. The installation was not completed before normal plant operation. The motive force (instrument air) and controller for the bypass valve were not installed. The design of the bypass valve was new to plant operators, and they had not received training on the operation of the valve. Craftsmen left the bypass valve in the open position with the manual mechanism not engaged. Operators later locked the handwheel in the open position, but did not engage the manual mechanism. This left the valve in a position where packing and actuator piston friction were the only things holding the valve open. Following a reactor trip, normal feedwater flow oscillations caused the valve to close and resulted in a loss of condensate and feedwater.

NRC Inspection Report No. 50-266/01-17; 50-301/01-17: An issue that arose during this inspection became a red finding (EA-02-031). Specifically, activities affecting quality were not prescribed by documented instructions, procedures, or drawings of a type appropriate to the circumstances. In particular, the emergency operating procedures (EOPs) did not provide adequate operator instructions to verify that AFW pump minimum flow recirculation valves were open while controlling AFW flow upon low IA header pressure. Low header pressure would cause the AFW pump minimum flow recirculation valves to fail closed — a significant condition adverse to quality that resulted in a potential failure of the AFW pumps as a result of blocking the discharge flow path. From at least 1997 through 2001, the licensee failed to promptly identify and correct this condition. The licensee's response to Generic Letter 88-14 could also have identified and addressed the AFW vulnerability associated with loss of IA.

EA-02-118 for Docket No. 50-456: The NRC issued a white finding and violation for the licensee's failure to correct and prevent recurrence of pressurizer PORV air accumulator check valve leak-through — a significant condition adverse to quality. Specifically, pressurizer relief valves failed to meet testing acceptance criteria in April 1991, October 1992, April 1994, January 1995, October 1995, October 1998, and September 2001. This resulted in several extended periods during which the unit was operated in a condition where the pressurizer PORVs may not have been able to perform their intended safety function of opening following events that resulted in isolation of IA to the containment or loss of the service air compressors.

EA-96-070 for Docket Nos. 50-456 and 50-457: The NRC issued a violation and a $100,000 civil penalty for a Severity Level III problem involving configuration control and corrective actions. One of the cited examples was the service air system being cross-connected to a water system on April 24, 1996.

EA-96-299 for Docket No. 50-309: The NRC issued violations for numerous problems. One Severity Level III problem related to corrective actions involved the IAS. Specifically, a loss of non safety-related IA could cause the air-operated dampers in the ducts of the containment spray building (CSB) fans to fail shut, rendering the fans incapable of performing their safety function of providing ventilation to the low-pressure safety injection (LPSI) and containment spray pumps and heat exchangers area. Without adequate ventilation, the LPSI and containment spray pump motors could fail as a result of overheating. The licensee identified this potential to lose safety-related CSB fans in 1991, but did not correct the problem until August 3, 1996.

EA-96-034 for Docket Nos. 50-245, 50-336, and 50-423: The NRC issued violations and imposed a $2,100,000 civil penalty for Severity Level II and III problems. One Severity Level II problem involved errors in design-basis documents. One example was that the licensee had inadequate provisions to ensure that design documents specify appropriate quality standards and that deviations from such standards are controlled. Specifically, the licensee procured filter regulators as nonsafety-related components, and installed them upstream of 48 safety-related SOVs to limit the differential pressure on the SOVs in accordance with design specifications. As a result, the licensee installed 48 SOVs that could be subject to air pressure in excess of the component designed maximum operating pressure differential if there was a failure of the nonsafety-related air regulator located upstream of each SOV.

Another example of this same Severity Level II problem was that from at least February 14, 1991, until 1996, the facility was not as described in the updated final safety analysis report (UFSAR), in that the Unit 1 diesel starting air receiver discharge check valve internals were removed, which defeated the capability for each air receiver to provide three independent cold diesel engine starts. The UFSAR was not complete and accurate in all material respects, in that it did not reflect this change. Prior to December 31, 1996, the EDG starting air system was configured in a manner different than described in the UFSAR, in that (1) while the ability to start the diesel engine three times at 250 psig without recharging the receivers was successfully demonstrated in the preoperational test, no supporting documentation was found that provided reasonable assurance that the receivers would contain sufficient inventory for three starts when the air receiver pressure is as low as 220 psig; (2) both compressors do not simultaneously receive a start signal; and (3) while the alternating current (AC) compressor starts at 225 psig, the direct current (DC) compressor starts at 220 psig.

A third example of this same Severity Level II problem at Unit 3 was that UFSAR Table 6.2-65, "Containment Penetration," identified the AFW flow control valves (FCVs) as containment isolation valves and indicated that they were motor-operated and fail "as is." UFSAR Section 6.2.4, "Containment Isolation System," stated that "all air- and solenoid-operated containment isolation valves fail in the closed position." Contrary to the above, as of May 22, 1996, the AFW FCVs, as originally installed, were solenoid-operated and failed "open," which constituted a change in the facility as described in the UFSAR. No evaluation existed to determine that the change did not constitute an unresolved safety question.

EA-97-007 for Docket No. 50-220: The NRC issued violations were issued for a Severity Level III problem and proposed a $50,000 civil penalty. One of the violations involved the Maintenance Rule (10CFR50.65) and failure to monitor the IAS isolation function, which prevents failures in the nonsafety-related portions of the system from affecting the safety-related portions of the system.

EA-96-370 for Docket Nos. 50-277 and 50-278: The NRC issued a Severity Level III violation under the Maintenance Rule for failure to monitor the performance or condition of numerous systems and components against established goals. One such system was the safety-grade instrument gas system.

EA-03-057 for Docket Nos. 50-266 and 50-301: The NRC issued a notice of violation associated with a red finding for Unit 2 (this was a yellow finding for Unit 1). The violation involved design control measures at the station. One example cited was the licensee's failure to correctly translate the AFW system design-basis power supply requirements into a modification package for the safety classification upgrade of the air-operated flow control valve (AOV) in each of the four recirculation lines. Specifically, the licensee did not ensure that the upgraded safety design relied only on a safety-related power source for a relay associated with the AOVs. Instead, the AFW system relied on a single train of nonsafety-related power supply for all trains of the AOV relays. Consequently, a common mode failure could have occurred during a loss of the nonsafety-related power supply.

EA-03-059 for Docket Nos. 50-266 and 50-301: The NRC issued a notice of violation associated with a previously identified red finding (see EA-02-031 below). This violation involved corrective actions associated with the AFW system. Specifically, the licensee failed to identify potential common mode failures that existed involving power supplies to the recirculation line air-operated valve and other system components. In addition, the licensee's corrective actions for the potential common mode failure associated with a loss of IA did not preclude repetition. Specifically, the licensee's corrective actions, to upgrade the safety function of the air-operated recirculation valve, failed to ensure that successful operation of the recirculation line air-operated valve was dependent only on safety-related support systems. Following the corrective actions, successful operation of the valve was still dependent upon nonsafety-related power to an interposing relay.

EA-02-031 for Docket Nos. 50-266 and 50-301: The NRC issued a notice of violation associated with a red finding involving the AFW system. Activities affecting quality were not prescribed by documented instructions, procedures, or drawings, of a type appropriate to the circumstances. Specifically, the Unit 1 and 2 EOPs for a reactor trip did not provide adequate operator instructions to verify that the AFW pump minimum flow recirculation valves were open while controlling AFW flow upon low IA header pressure. Low header pressure would cause the AFW pump minimum flow recirculation valves to fail closed — a significant condition adverse to quality that resulted in potential failure of the AFW pumps as a result of blocking the discharge flow path. From at least 1997 through 2001, the licensee failed to promptly identify and correct this condition adverse to quality. Prior opportunities to identify this failure mode arose in October 1997 when the safety function of the minimum flow recirculation valves was considered in response to a condition report, and again in March 1997 when the licensee identified a failure mode of the AFW system attributable to the loss of IA as discussed in an LER.

EA-98-478 for Docket No. 50-458: The NRC proposed a $55,000 civil penalty and issued two Severity Level III violations for design control measures and corrective actions involving the IA supply for the EDGs. Since 1985, design control measures did not adequately provide for verifying or checking (through the performance of design reviews, use of alternative or simplified calculational methods, or performance of testing) that the safety-related diesel generator control air instrument and controls system remained functional during accident conditions. Specifically, design control measures did not ensure that the system was provided with a long-term supply of safety-related pressurized air, which was necessary for the continued operation of the diesel generators in response to an extended loss of offsite power (i.e., the air compressors were nonsafety-related and were not powered by a safety-related bus). At less than 120 psig, the non-essential diesel generator trips would no longer be bypassed, and at less than 45 psig, the diesel generators would automatically shut down. As a result, the Division I and II diesel generators were not operable while in Modes 1, 2, and 3 during this time period because the control air instrument and controls subsystems were not operable.

From 1985 until about June 1998, a significant condition adverse to quality existed related to the Division I and II diesel generator control air instrument and controls subsystems, and the cause of the condition was not determined, and adequate corrective action was not taken throughout this time. Since 1990, licensee staff knew that diesel generator control air instrument and controls subsystems were not provided with a long-term source of safety-related pressurized air to ensure that the nonessential diesel generator trips would remain bypassed during a loss of offsite power. Although the licensee changed its procedures in 1990 to require operators to install nonsafety-related air bottles as an alternative air source, the licensee did not properly evaluate the acceptability of relying on this operator action (in lieu of automatic action) against its design-basis description in the safety analysis report, and did not fully demonstrate the ability to accomplish the manual actions until 1998. Further, the failure to identify the significant condition adverse to quality continued until 1998, and the licensee did not document the cause of the condition and the corrective action taken and did not report them to appropriate levels of management.

EA-97-055 for Docket Nos. 50-280 and 50-281: The NRC issued a violation under the Maintenance Rule and proposed a $55,000 civil penalty for a Severity Level III problem. One example cited was that the licensee failed to demonstrate that the performance of the IA compressor had been effectively controlled through the performance of appropriate preventive maintenance. Specifically, the licensee failed to establish any measure to evaluate the appropriateness of the performance of preventive maintenance on the IA compressor before placing it under Section (a)(2) of the Maintenance Rule.

EA-97-531 for Docket No. 50-271: The NRC issued a Severity Level IV violation for the licensee's failure to correctly select equipment in a subsystem essential to the safety-related function of the EDGs. Specifically, air to the solenoid valves that operated the EDG service water cooling FCVs was supplied from a nonsafety-related pressure regulator. Failure of the pressure regulator could have resulted in a malfunction of the solenoid valve, which could have prevented the FCVs from opening. The failure of the flow control valve could cause a loss of all service water to the EDGs, which would prevent their operation.

<u>Findings from Inspection and Escalated Enforcement Actions Involving Air Systems</u>

The staff's analysis revealed the following findings concerning the inspection and escalated enforcement actions involving air systems and components:

- The findings or actions involved weaknesses or violations (in order of decreasing frequency) in design (including engineering, modification, and configuration control), problem identification and corrective actions, maintenance, and training or procedures. Many of the findings or violations involved weaknesses or violations in more than one area.

- The most risk-significant findings under the current oversight process involved the AFW system (three red findings), condensate and feedwater (one white finding), and PORVs (one white finding). Significant findings and actions under the old oversight process involved EDGs (three actions), and low-pressure coolant injection and containment spray (one action).

- The significance and long-standing nature of many of the conditions leading to inspection findings and escalated enforcement actions, especially after the issuance of Generic Letter 88-14, suggest that initial licensee and NRC activities to identify and resolve air system problems had only limited effectiveness.

3.4 Generic Issue Process and Generic Letter Closeout

3.4.1 Generic Issue Process

As discussed in Section 2.2, the NRC staff prioritized Generic Issue 43 in 1983, and recommended dropping it from further consideration. However, following the publication of the priority evaluation in November 1983, the staff received comments from ACRS and AEOD. Rather than agreeing with the staff's recommendation, the ACRS and AEOD comments recommended that the staff should broaden the issue to include all causes of air system unavailability, instead of confining it with the restrictive limits that were previously imposed in Generic Issue 43. Apparently, Generic Issue 43 would have been dropped at this point absent the ACRS and AEOD recommendations.

After the completion of the extensive AEOD case study in 1987, the NRC staff reevaluated, broadened, and retitled Generic Issue 43, and subsequently assigned the issue a high priority ranking based on its value/impact score. As noted in Section 2.2, there were significant recognized limitations in the analyses used to assign this high priority ranking. However, operating experience in LERs revealed numerous additional instances in which a high degree of risk sensitivity was apparent. The staff then issued Generic Letter 88-14 on August 8, 1988 (Ref. 6), to alert licensees and industry stakeholders to the broader issue concerning Instrument air supply system problems affecting safety-related equipment. The staff considered Generic Issue 43 resolved with the issuance of Generic Letter 88-14.

The following improvement opportunities existed in this historic example:

- Generic Issue 43 arose from operating experience and was nearly dropped with no action taken.

- To resolve Generic Issue 43, the staff issued Generic Letter 88-14. Numerous events, catalogued in Reference 4, occurred while Generic Issue 43 was being processed.

- The staff considered Generic Issue 43 resolved with the issuance of Generic Letter 88-14. Current practice is to only consider a generic issue resolved after licensees or certificate holders implement requirements or adhere to the guidance NRC disseminated for that issue and the NRC verifies its implementation.

3.4.2 Generic Letter Closeout

Licensees made written submittals in responses to Generic Letter 88-14, and the NRC staff reviewed and subsequently approved these responses. The related correspondence reveals that there were frequently several iterations (i.e., teleconferences, meetings, or requests for additional information) before the staff found the licensees responses acceptable. The staff typically sent short letters to the licensees informing them that their submittals were acceptable and that post-implementation audit inspections would be conducted to verify the adequacy of licensees' efforts.

A detailed review of findings and related enforcement actions from the implementation audit inspections was not conducted for this assessment because retrieving and searching relevant documents reports from 1989 to 1991 is impractical with the limited ADAMS search capability of archived Nuclear Documents System (NUDOCS) correspondence. Nonetheless, it is clear from LERs (see Sections 3.1 and 3.2) that implementation audit inspections identified numerous reportable conditions involving air systems. The limited sampling of correspondence also shows that the NRC staff inspected, in detail, the resolution of Generic Letter 88-14 and identified instances in which licensees did not appear to meet commitments. Reference 10 provides one example.

4. FINDINGS AND CONCLUSIONS

The assessment developed findings:

- From other studies of operating experience (Section 3.1.4):

  — Instrument air system and component performance has significantly improved since 1987.

  — Prompt recovery from a major loss of IA is typical at PWRs. All but one of the events that occurred since the issuance of Generic Letter 88-14 were recovered within 4 minutes.

- From the 26 LERs that involved air systems and referenced Generic Letter 88-14 since 1988 (Section 3.2.1):

  — Issuance of Generic Letter 88-14 and targeted NRC inspections led to the identification of problems in safety-related systems and components by both the NRC and licensees.

  — The LERs reported a wide variety of problems and programmatic issues, and a single LER often contained several related issues.

  — About 80 percent (21 of 26) of these LERs were reported before 1991. Half of the LERs described previous missed opportunities to identify the reportable condition, and three of these LERs described the missed opportunities as including licensee activities associated with Generic Letter 88-14.

  — About 65 percent of the conditions reported in the LERs were identified by the licensee, about 23 percent were self-revealing, and about 12 percent were identified by the NRC.

- From the 58 LERs that involved air system design conditions and potential or actual degradation of at least one train of a safety system since 1988 (Section 3.2.2):

  — Issuance of Generic Letter 88-14 led to the identification and correction of problems in many safety-related systems and components.

  — The major safety systems involved (in order of decreasing frequency) were emergency power, injection and cooling systems, containment isolation, AFW, service water, containment spray, and component cooling water.

  — About 28 percent of these LERs were reported before 1991, and some licensees reported LERs before the generic letter was issued. About 72 percent of these LERs were reported in 1991 or later.

  — Most of these LERs involved long-standing conditions, with two-thirds (or more) existing since initial licensing and construction.

  — About 80 percent of the conditions reported in the LERs were identified by the licensee, about 10 percent were self-revealing, and about 10 percent were identified by the NRC. About half of the NRC-identified conditions were in the 1988–1990 time frame, when the NRC was inspecting air systems and reviewing licensee actions in response to Generic Letter 88-14.

- From the inspection findings and escalated enforcement actions involving air systems and components (Section 3.3):

  — The findings or actions (in order of decreasing frequency) involved weaknesses or violations in design (including engineering, modification, and configuration control), problem identification and corrective actions, maintenance, and training or procedures. Many of the findings or violations involved weaknesses or violations in more than one area.

  — The most risk-significant findings under the current oversight process involved the AFW system (three red findings), condensate and feedwater (one white finding), and PORVs (one white finding). Significant findings and actions under the former oversight process involved EDGs (three actions) and low-pressure coolant injection and containment spray (one action).

On the basis of its assessment, the staff reached the following conclusions:

- Licensee and agency activities, such as the Maintenance Rule, Generic Letter 88-14, design-basis reconstitution, and others, have significantly improved air system and component performance and, thereby, resulted in improved reactor safety.

- Based on PWR data, major losses of instrument air are now infrequent, and prompt recovery from such losses is typical, indicating that the actions requested by Generic Letter 88-14 have contributed to improved reactor safety.

- Issuance of Generic Letter 88-14 and targeted NRC inspections led to the identification and resolution of air system design issues impacting safety-related systems and components, and resulted in improved reactor safety.

- As evidenced by the ongoing discovery and correction of air system issues, licensee programs and NRC oversight activities provide assurance that the NRC and its licensees are effectively maintaining reactor safety in this area.

5. REFERENCES

1. Memorandum from C. Michelson, NRC, to H.R. Denton, NRC, and V. Stello, Jr., NRC, "Immediate Action Memorandum: Common Cause Failure Potential at Rancho Seco — Desiccant Contamination of Air Lines," September 15, 1981 [NUDOCS Accession No. 8109280036].

2. Memorandum from H.R. Denton, NRC, to C. Michelson, NRC, "AEOD Immediate Action Memorandum on Contamination of Instrument Air Lines at Rancho Seco," October 28, 1981 [NUDOCS Accession No. 8111300391].

3. NUREG-0933, "A Prioritization of Generic Safety Issues," August 2004 [available at http://www.nrc.gov/reading-rm/doc-collections/nuregs/staff/sr0933/].

4. NUREG-1275, Vol. 2, "Operating Experience Feedback Report — Air System Problems," December 1987 [NUDOCS Accession No. 8801070069].

5. NRC Information Notice 87-28, "Air System Problems at U.S. Light Water Reactors," Supplement 1, December 28, 1987 [available at http://www.nrc.gov/reading-rm/doc-collections/gen-comm/info-notices/1987/in87028.html]

6. NRC Generic Letter 88-14, "Instrument Air Supply System Problems Affecting Safety-Related Equipment," August 8, 1988 [NUDOCS Accession No. 8808120294].

7. NRC Information Notice 2002-29, "Recent Design Problems in Safety Functions of Pneumatic Systems," October 15, 2002 [available at http://www.nrc.gov/reading-rm/doc-collections/gen-comm/info-notices/2002/in200229.pdf]

8. "Rates of Initiating Events at U.S. Nuclear Plants: 1988–2003" (update to NUREG/CR-5750), Idaho National Engineering and Environmental Laboratory, September 2004 [available at http://nrcoe.inel.gov/results/index.cfm?fuseaction=InitEvent.showMenu].

9. "Component Performance Study of Air-Operated Valves 1987–2003" (update to NUREG-1715, Volume 3), Idaho National Engineering and Environmental Laboratory, September 2004 [available at http://nrcoe.inel.gov/results/index.cfm?fuseaction=CompPerf.showMenu].

10. NRC Inspection Report Nos. 50-327/90-025 and 50-328/90-25, November 1, 1990 [NUDOCS Accession No. 9012050094].

NRC FORM 335
(2-89)
NRCM 1102,
3201, 3202

U.S. NUCLEAR REGULATORY COMMISSION

BIBLIOGRAPHIC DATA SHEET

(See instructions on the reverse)

1. REPORT NUMBER
(Assigned by NRC, Add Vol., Supp., Rev.,
and Addendum Numbers, if any.)

NUREG-1837

2. TITLE AND SUBTITLE

Regulatory Effectiveness Assessment of Generic Issue 43 and Generic Letter 88-14

3. DATE REPORT PUBLISHED

MONTH	YEAR
October	2005

4. FIN OR GRANT NUMBER

5. AUTHOR(S)

John V. Kauffman

6. TYPE OF REPORT

Technical

7. PERIOD COVERED *(Inclusive Dates)*

8. PERFORMING ORGANIZATION - NAME AND ADDRESS *(If NRC, provide Division, Office or Region, U.S. Nuclear Regulatory Commission, and mailing address; if contractor, provide name and mailing address.)*

Division of Systems Analysis and Regulatory Effectiveness
Office of Nuclear Regulatory Research
U.S. Nuclear Regulatory Commission
Washington, DC 20555-0001

9. SPONSORING ORGANIZATION - NAME AND ADDRESS *(If NRC, type "Same as above"; if contractor, provide NRC Division, Office or Region, U.S. Nuclear Regulatory Commission, and mailing address.)*

Same as above

10. SUPPLEMENTARY NOTES
John V. Kauffman, NRC Project Manager

11. ABSTRACT *(200 words or less)*

This report documents an assessment of Generic Issue 43, "Contamination of Instrument Air Lines," and Generic Letter 88-14, "Instrument Air Supply System Problems Affecting Safety-Related Equipment." This assessment is part of an ongoing initiative through which the U.S. Nuclear Regulatory Commission (NRC), Office of Nuclear Regulatory Research, is reviewing selected agency regulations and decisions to determine whether they are achieving the desired results. For this assessment, the staff compared expectations with outcomes. Whenever outcomes fell short of expectations, the staff attempted to identify ways to enhance the effectiveness, efficiency, and realism of the NRC's regulatory process.

On the basis of its assessment, the staff concluded that licensee and agency activities, such as the Maintenance Rule, Generic Letter 88-14, design-basis reconstitution, and others, have significantly improved air system and component performance and, thereby, resulted in improved reactor safety. Moreover, issuance of Generic Letter 88-14 and targeted NRC inspections led to the identification and resolution of air system design issues impacting safety-related systems and components, again resulting in improved reactor safety. As a result, based on data for pressurized-water reactors, major losses of instrument air are now infrequent, and prompt recovery from such losses is typical, which supports the staff's conclusion that reactor safety has improved. In addition, as evidenced by the ongoing discovery and correction of air system issues, licensee programs and NRC oversight activities provide assurance that the NRC and its licensees are effectively maintaining reactor safety in this area.

12. KEY WORDS/DESCRIPTORS *(List words or phrases that will assist researchers in locating the report.)*

air-operated valve
air operated valve
AOV
air system
common-cause failure
Generic Issue 43
Generic Letter 88-14
operating experience
pneumatic system
regulatory effectiveness

13. AVAILABILITY STATEMENT

unlimited

14. SECURITY CLASSIFICATION

(This Page)

unclassified

(This Report)

unclassified

15. NUMBER OF PAGES

16. PRICE

NRC FORM 335 (2-89)

www.ingramcontent.com/pod-product-compliance
Lightning Source LLC
Chambersburg PA
CBHW081403170526
45166CB00010B/3192